Contents

Dedication

This book is devoted to all the enthusiastic people who work endlessly in the fields of artificial intelligence and software testing.

To individuals who aim for excellence in guaranteeing AI-driven systems' dependability, security, and moral norms. We are all inspired by your passion for expanding the boundaries of AI technology and your devotion to the craft of testing.

To the instructors and mentors who kindly impart their wisdom, directing and molding the upcoming cohort of artificial intelligence tests. In order to develop future professionals in this ever-evolving industry, your wisdom and guidance are crucial.

To the trailblazers and innovators who are expanding the realm of AI testing possibilities. Your inventiveness and foresight provide doors for groundbreaking discoveries that benefit society.

To the innovators and trailblazers pushing the boundaries of what is possible in AI testing. Your creativity and vision pave the way for groundbreaking advancements that benefit society as a whole.

Finally, we would like to express our gratitude to our families and loved ones, whose continual support and comprehension have been a source of strength for us during this trip. We are incredibly grateful for your support and faith in our efforts.

This book is devoted to each of you, recognizing your commitment, love, and contributions to the field of artificial intelligence testing.

How to Read this Book

Welcome to *AI Testing Demystified A Comprehensive Handbook.* Consider the following tips to get the most out of this book and enhance your learning experience:

1. Sequential Approach: The book is organized in a logical sequence, covering key themes ranging from the evolution of software testing to new trends in AI testing. Reading the chapters in order will give you a thorough understanding of the subject.

2. Skimming and Focus Reading: Each chapter begins with an overview to help you understand the essential themes. Skim over the content at first to get a sense of the subjects covered. For a deeper dig, concentrate on topics relevant to your learning objectives.

3. Hands-on Exercises: Hands-on examples and practical exercises are included in chapters for interactive learning. Engage actively with these sections, since they present real-world experiences to help you comprehend.

4. Case Studies and Examples: The real-world case studies and examples provided provide practical insights into the application of AI testing principles. Consider these examples to connect theory to practice.

5. Glossary of words in the Appendices: The glossary of words in the appendices serves as a quick reference for unfamiliar terms encountered throughout the book. Use the list of resources and references for additional inquiry and learning outside the scope of the book.

6. Tailoring to Your Needs: Tailor your reading experience to your interests and goals. Whether you are studying for certification, honing specific abilities, or gaining a larger grasp of AI testing, tailor your reading strategy accordingly.

7. Review and Revisit: Go over essential concepts again and revisit parts as needed. Use the book as a reference guide for continuous AI testing concepts learning and reinforcement.

We hope these suggestions assist you in properly navigating the book and making the most of the extensive knowledge given therein.

Part 1: Introduction to AI Testing

Chapter 1: The Evolution of Software Testing

In this chapter, we will take a journey through the history of software testing, exploring how it has evolved. We will also delve into the exciting emergence of artificial intelligence (AI) in the field of testing.

Historical perspective

<table>
<tr>
<td>

Early Computing Era (1940s):

Manual and Tedious Testing: Enormous machines required manual validation of each code line.
Challenges: Tedious, error-prone, and time-consuming process.

</td>
<td>

Structured Testing Emergence (1960s - 1970s):

Structured Methodologies: Introduction of formal techniques and standardized processes.
Shift in Approach: Transition from ad-hoc to systematic and organized testing approaches.

</td>
</tr>
<tr>
<td>

PC Era (1980s):

Complexity and Opportunities: Rise of personal computers led to more complex software.
Automation Emergence: Need for automated testing tools for efficiency.
Test Automation: Enabled quicker and more accurate test case execution.

</td>
<td>

Agile and DevOps Era (21st Century):

Continuous Testing Emphasis: Agile and DevOps introduced continuous testing throughout the development lifecycle.
Automated and Continuous Testing: Vital to match fast-paced development cycles.

</td>
</tr>
</table>

Software testing has come a long way since the earliest days of computing.

In the early 1940s, when computers were enormous machines that filled entire rooms, testing was a manual and time-consuming process. Testers had to meticulously examine and validate each line of code, looking for errors and glitches. It was a tedious, error-prone, and time-consuming task.

As technology advanced, so did the need for more efficient testing methods. In the 1960s and 1970s, structured testing methodologies began to emerge. Testers started using formal techniques and standardized processes to improve the quality of software. This marked a significant shift from ad-hoc testing to systematic and organized approaches.

The advent of personal computers in the 1980s brought new challenges and opportunities for software testing. As software became more complex and widespread, the need for automated testing tools became evident. Test automation started gaining popularity, enabling testers to run test cases more quickly and accurately.

With the turn of the 21st century, we entered the era of Agile and DevOps methodologies, which emphasized the importance of continuous testing throughout the software development lifecycle. This shift increased the need for automated and continuous testing to keep up with the fast-paced development cycles.

AI: The Game-Changer in Software Testing

In the fast-paced world of software development, testing has always been the unsung hero, ensuring that the products we rely on are bug-free and user-friendly. However, as software complexity continues to grow, traditional testing methods are struggling to keep up. Enter artificial intelligence (AI), a revolutionary technology that is transforming the testing landscape.

AI is like having a super powered assistant in your testing team, capable of automating tasks, analyzing data, and predicting potential problems. It is not just about replacing manual work; it is about augmenting human capabilities, making testing more efficient, accurate, and proactive.

Imagine test scripts being generated automatically, saving testers hours of tedious work. Envision predictive analytics identifying potential issues before they become major problems, preventing costly delays and customer frustration. Picture test data that mirrors real-world scenarios, ensuring that your software is ready for any challenge.

AI is breathing new life into software testing, making it more agile, responsive, and aligned with the demands of modern development. It is like having a secret weapon in your arsenal, giving you the edge to create software that is not just bug-free but also exceptional.

Here are some **real-world examples** of how AI is transforming software testing:

Test Automation:

AI-powered test scripts
Reduced manual effort
Improved test coverage

Predictive Analysis:

Analysis of historical test data
Identification of potential issues
Prioritization of testing efforts

Test Data Generation:

Simulation of real-world scenarios
Comprehensive test coverage
Identification of edge cases

Defect Prediction:

Identification of patterns in code
Early defect prevention
Reduced development costs

Natural Language Processing (NLP):

Understanding and processing natural language requirements
Streamlined testing process
Improved communication between testers and developers

AI is not just a trend; it is a revolution in software testing. It is empowering testers to do their jobs better, faster, and more efficiently, leading to better software for everyone. Therefore, embrace the future of testing and let AI be your guide.

Part 2: AI Fundamentals

Chapter 2: Introduction to Artificial Intelligence

What is AI?

Artificial Intelligence commonly referred to as AI, stands at the forefront of computer science, striving to create systems capable of emulating human intelligence.

It is a fascinating realm where machines delve into problem-solving, learning through experiences, comprehending natural language, and making decisions, activities traditionally associated with human cognition. In this exciting field, AI systems manifest in various forms, ranging from straightforward rule-based programs to intricate neural networks mimicking the complexities of the human brain.

Narrow or Weak AI

Specialized Intelligence: Tailored for specific tasks or domains.
Examples:
Virtual Assistants (e.g., Siri, Alexa): Assist with specific commands or queries.
Recommendation Engines (e.g., Netflix): Suggest personalized content based on preferences.
Characteristics:
Excel within defined tasks but lack broad human-like intellect.
Application-specific intelligence rather than general cognitive abilities.

General or Strong AI

Human-like Intelligence: Hypothetical level where AI mirrors human cognition.
Potential Capabilities:
Ability to perform any cognitive task akin to humans.
Comprehensive understanding, reasoning, and learning across diverse domains.
Status:
Frontier yet to be fully realized but holds immense potential and

They come in two main categories:

In this quest for AI evolution, the boundary between what is possible and what is yet to come remains a tantalizing area of exploration.

AI in software testing

AI has been making waves in the field of software testing, transforming how testing is conducted and optimizing its effectiveness.

Test Automation:

AI-Powered Tools: Automate test case generation, execution, and reporting.
Benefits:
Speeds up testing processes.
Reduces manual testing efforts.
Example:
Selenium with AI capabilities adapts test scripts based on application behavior changes.

Defect Prediction:

AI Analysis: Predicts likely defect-prone areas by analyzing code and historical data.
Benefits:
Guides testers to focus on critical areas.
Example:
Identifying patterns between code modules and defect occurrence using historical project data.

Natural Language Processing (NLP):

Understanding Natural Language Requirements: Valuable for applications with user-generated content.
Benefits:
Facilitates comprehension of natural language interfaces.
Example:
AI-driven NLP tools extract test scenarios from plain language descriptions.

Test Data Generation:

Synthetic Test Data Creation: Generates realistic test data for comprehensive testing.
Benefits:
Essential for extensive testing coverage, particularly with large datasets.
Example:
AI creates synthetic customer profiles for banking applications to test various financial scenarios.

Here are some key ways in which AI is utilized in software testing:

AI testing techniques

AI testing techniques leverage artificial intelligence to improve the quality and efficiency of software testing. Here are some key AI testing techniques:

- **Machine Learning-Based Testing:** Machine-learning algorithms can be used to identify patterns in testing data, automate test case generation, and predict potential defects.
- **Visual AI Testing:** AI can analyze and compare screenshots of an application's user interface to detect visual defects, ensuring consistent user experiences.
- **AI-Powered Test Case Prioritization**: AI can determine the order in which test cases should be executed based on the likelihood of defects and the impact on the application.
- **AI-Driven Test Maintenance:** AI can automatically update test scripts when the application changes, reducing the maintenance burden.

Machine Learning-Powered Testing

Supervised learning for defect prediction

Unsupervised learning for pattern discovery

Reinforcement learning for optimal test case selection

Visual AI Testing

Image recognition algorithms for defect identification

Pixel-level comparison for UI consistency

Anomaly detection for unexpected UI elements

AI-Driven Test Case Prioritization

Risk-based prioritization for critical defects

Impact-based prioritization for user-facing features

Dynamic prioritization for real-time testing scenarios

AI-Powered Test Maintenance

Natural language processing for test case understanding

Change detection for test script adaptation

Self-healing mechanisms for proactive maintenance

In the following chapters of this book, we will delve deeper into these AI testing techniques and provide practical examples to help you understand how AI can be applied to enhance your software testing processes. Whether you are a beginner or an experienced tester, embracing AI in testing can make your work more efficient and effective.

Chapter 3: Machine Learning and AI Models

In this chapter, we will explore the fundamental concepts of machine learning and delve into the different types of AI models. Additionally, we will discuss the lifecycle of AI models, from their creation to deployment and maintenance.

Basics of machine learning

Machine learning is a subfield of artificial intelligence that focuses on creating algorithms and models that enable computers to learn from and make predictions or decisions based on data. Let us break down some key concepts:

1. **Data**: Machine learning relies on data, which can be any kind of information, such as text, images, or numbers. Data is used to train models and make predictions.
2. **Features**: Features are the specific characteristics or attributes within the data that the machine-learning model uses to make predictions. For example, in image recognition, features could be pixel values.
3. **Training**: During the training phase, a machine-learning model learns patterns and relationships within the data. It adjusts its internal parameters to make accurate predictions.
4. **Testing and Validation:** Once trained, models are tested and validated using separate datasets to ensure they can make accurate predictions on new, unseen data.
5. **Prediction/Inference:** After training, models are used to make predictions or inferences on new data. For example, a recommendation system predicts which products a user might be interested in.

Data

Structured data
Tabular format, easily processable
Unstructured data
Text, images, audio, video
Data cleaning and preprocessing

Features

Domain knowledge
Understanding the data and context
Feature selection
Choosing relevant and informative features
Feature transformation
Encoding and scaling features

Training

Supervised learning
Learning from labeled data
Unsupervised learning
Discovering patterns in unlabeled data
Reinforcement learning
Learning through trial and error

Testing and Validation:

Holdout method
Splitting data into training and testing sets
Cross-validation
Evaluating model performance on multiple
    subsets
Error metrics
Measuring accuracy, precision, recall

Prediction/ Inference:

Classification
Assigning labels to new data points
Regression
Predicting continuous numerical values
Clustering
Grouping similar data points together

Types of AI models

AI models come in various types, each designed for different tasks. Here are some common types of AI models:

1. **Supervised Learning:** In supervised learning, models are trained on labeled data, where both input and output are known. For example, a spam email filter is trained on labeled emails (spam or not spam).
2. **Unsupervised Learning**: Unsupervised learning models work with unlabeled data and aim to discover patterns, clusters, or structures within the data. For example, clustering customer data to identify market segments.
3. **Reinforcement Learning:** Reinforcement learning involves agents learning to make decisions through trial and error. They receive rewards for good actions and punishments for bad ones. Examples include training robots to perform tasks.
4. **Deep Learning:** Deep learning is a subset of machine learning that uses artificial neural networks inspired by the human brain. It is particularly powerful for tasks like image and speech recognition.
5. **Natural Language Processing (NLP):** NLP models are specialized in understanding and generating human language. They are used for tasks like text classification, sentiment analysis, and machine translation.

Supervised Learning	Models trained on labeled data where input and output are known. Spam email filter trained on labeled emails (spam or not spam).
Unsupervised Learning	Models work with unlabeled data to discover patterns or clusters. Customer data clustering for market segment identification.
Reinforcement Learning	Agents learn through trial and error, receiving rewards and punishments. Training robots to perform tasks based on feedback.
Deep Learning	Subset using artificial neural networks, powerful for image and speech recognition. Image recognition systems using deep learning algorithms.
Natural Language Processing (NLP)	Specialized models for understanding and generating human language. Text classification, sentiment analysis, machine translation.

AI model lifecycle

The lifecycle of an AI model involves several stages, from conception to deployment and ongoing maintenance:

<table>
<tr>
<td>

Problem Definition:

Clearly define the problem the model aims to solve.

Establish goals and objectives for the AI model.

</td>
<td>

Data Collection:

Gather relevant and high-quality data required for training the AI model.

Data quality significantly impacts model accuracy.

</td>
</tr>
<tr>
<td>

Data Preprocessing:

Prepare and clean the collected data.

Handle missing values, outliers, and ensure data is formatted correctly for model training.

</td>
<td>

Model Selection and Training:

Choose an appropriate model type and train it using the prepared dataset.

Model selection impacts the model's performance.

</td>
</tr>
<tr>
<td>

Validation and Testing:

Evaluate the trained model's performance using separate datasets.

Ensure it can make accurate predictions before deployment.

</td>
<td>

Deployment:

Deploy the trained model into a production environment where it can be used for real-time predictions or tasks.

</td>
</tr>
<tr>
<td>

Monitoring and Maintenance:

Continuously monitor the deployed model's performance.

Retrain or update the model as needed to adapt to evolving data patterns and ensure sustained accuracy.

</td>
<td></td>
</tr>
</table>

Example:

Here's the content tailored for the development of a recommendation system used by an e-commerce platform:

Problem Definition:

- **Objective**: The goal is to suggest relevant products to users based on their preferences and behavior.
- **Data Collection:** Gather user behavior data, including browsing history, purchase records, items viewed, and interactions.

Data Preprocessing:

- **Data Preparation:** Clean and preprocess collected data, handling missing values, removing duplicates, and ensuring data quality.
- **Data Formatting:** Organize data into suitable formats for model training, considering features such as user preferences, item characteristics, and historical interactions.

Model Selection and Training:

- **Algorithm Selection:** Choose appropriate recommendation algorithms (e.g., collaborative filtering, content-based, or hybrid models).
- **Model Training:** Train the selected model(s) using the preprocessed data to learn patterns and generate recommendations.

Validation and Testing:

- **Evaluation Metrics:** Assess model accuracy and performance using validation datasets, measuring metrics like precision, recall, or mean average precision.
- **Testing Reliability:** Ensure the recommendation model provides relevant and diverse suggestions to users.

Deployment:

Integration: Deploy the trained recommendation model onto the e-commerce platform, enabling users to receive personalized product recommendations.

Real-time Recommendations: Enable real-time recommendation generation based on user interactions.

Monitoring and Adaptation:

- **Continuous Monitoring:** Monitor user interactions and feedback to evaluate the effectiveness of recommendations.
- **Adaptation and Updates:** Modify the recommendation system over time based on changing user preferences and behaviors.

This content outlines the essential steps involved in developing and deploying a recommendation system for an e-commerce platform. It covers key stages from problem definition to the continuous monitoring and adaptation of the recommendation model based on user interactions.

Understanding these fundamental concepts and the different types of AI models is essential for anyone involved in AI testing. In the subsequent chapters, we will explore how AI models are tested and evaluated in the context of software testing.

Chapter 4: Data in AI Testing

In this chapter, we will dive into the critical role of data in AI testing. We will explore the processes of data collection and preprocessing, discuss the types of datasets used for testing, and address the crucial aspects of data quality and privacy.

Data collection and preprocessing

Before AI models can be trained and tested, they need data. Data collection and preprocessing are the foundational steps in AI testing:

Data Collection

Definition: Gathering relevant data from diverse sources.

Variety of Data: Data can be in multiple forms: text, images, sensor readings, etc.

Example: Autonomous Vehicle Testing - Collecting data from cameras, lidar, and sensors for safe driving models.

Data Preprocessing

Preparing Raw Data: Raw data is often untidy and unstructured.

Tasks Involved: Cleaning, transforming, and organizing data.

Normalization: Tasks include handling missing values, removing outliers, and normalizing data.

Datasets for testing

AI testing relies on datasets to assess the performance and accuracy of models. Datasets are collections of data that have been labeled or categorized for specific tasks:

Training Datasets
> *Purpose:* Used to train AI models.
> *Contents:* Examples with input data and expected output.
> *Example:* Spam Email Filter - Dataset labeled "spam" or "not spam."

Validation Datasets
> *Role during Training:* Fine-tuning and assessing model performance.
> *Ensuring Generalization:* Helps models generalize to unseen data.

Testing Datasets
> *Post-Training Evaluation:* Assess model performance after training.
> *Unseen Data:* Contains data the model has not encountered before.
> *Real-world Evaluation:* Measures model predictions in real-world scenarios.

Public Datasets
> *Availability:* Provided by various organizations and research groups.
> *Example:* ImageNet Dataset - Millions of labeled images for image recognition tasks.

Data quality and privacy

Data

Importance: Crucial for training accurate models.
Impact of Poor Data Quality: Can lead to biased or inaccurate predictions.
Ensuring Data Quality:
Data Cleaning: Removing inconsistencies, errors, or duplicates.
Validation: Verifying accuracy and reliability.
Representation: Data should represent the problem being solved.
Example - Medical Image Analysis:
Importance: Avoiding misdiagnoses in medical images.
Requirements: Clear, well-labeled, artifact-free images.

Data

Legal and Ethical Consideration: Protecting individuals' privacy and sensitive information.
Privacy Measures:
Anonymization: Masking or removing identifiable information.
Encryption: Securing data from unauthorized access.
Example - Healthcare:
Importance: Adhering to strict privacy regulations (e.g., HIPAA in the US).
Sensitive Data: AI models using medical data must comply with privacy laws.

Data quality and privacy are paramount in AI testing:

Balancing data quality and privacy is a common challenge in AI testing. Testers must ensure data is accurate and representative while respecting privacy and security concerns.

In the following chapters, we will explore specific AI testing techniques that rely on these datasets, such as test case generation and defect prediction. Understanding data in AI testing is fundamental because the quality and relevance of your data play a significant role in the success of AI testing projects.

Part 3: AI Testing Techniques

Chapter 5: Test Strategy for AI Applications

In this chapter, we will focus on developing a comprehensive test strategy for AI applications. We will address the unique challenges that AI testing poses, explore the concept of risk-based testing, and guide you through effective test planning for AI projects.

AI-specific test challenges

Testing AI applications introduces specific challenges that set it apart from traditional software testing:

Lack of Deterministic Outcomes:
- *Probabilistic Results:* Difficulty in defining a single "correct" output.
- *Example - Language Translation Model:* Offers multiple valid translations for a given input.

Data-Dependent Behavior:
- *Reliance on Training Data:* Quality and diversity significantly impact model behavior.
- *Risk of Bias:* Small or biased datasets can skew model behavior.
- *Example - Speech Recognition Model:* Struggles with accents not well-represented in training data.

Model Drift:
- *Degradation of Accuracy Over Time:* Real-world data differs from training data.
- *Adaptation Challenges:* Models may fail to adapt to changing scenarios.
- *Example - Recommendation System:* Fails to reflect changing user preferences and trends.

Black-Box Nature:
- *Complex Inner Workings:* Difficulty in understanding the model's operations.
- *Challenge in Failure Analysis:* Identifying root causes of model failures.
- *Implications:* Lack of transparency complicates troubleshooting.

Risk-based testing

Given the unique challenges of AI testing, a risk-based approach is crucial. Risk-based testing involves identifying and prioritizing the most critical aspects to focus testing efforts on.

Identify Risks:

Safety Concerns: Identify potential hazards in AI-driven systems.
Regulatory Compliance: Ensure adherence to industry regulations.
Critical Business Processes: Assess AI's impact on vital business operations.

Risk Assessment:

Impact Evaluation: Assess potential consequences of identified risks.
Likelihood Analysis: Evaluate the probability of risks occurring.

Testing Prioritization:

Focused Resource Allocation: Allocate more resources to high-impact, high-likelihood risks.
Thorough Testing: Ensure comprehensive testing coverage for critical areas.

Here is how to approach it:

Test planning for AI

Effective test planning is vital for successful AI testing:

Define Objectives:
Clearly state the objectives of your AI testing.
What are you trying to achieve?
Are you validating a model's accuracy, assessing its reliability, or ensuring its ethical behavior?

Ethical Considerations:
Ensure that your testing approach takes into account ethical and fairness concerns.
Test for bias in AI models to avoid discriminatory behavior.

Test Data Strategy:
Plan how you will source, preprocess, and manage test data.
Ensure it is representative of real-world scenarios and covers the most critical use cases.

Metrics and KPIs:
Define key performance indicators (KPIs) and metrics to assess the model's performance.
For example, in speech recognition, word error rate (WER) is a common metric.

Testing Environment:
Create a controlled testing environment to simulate real-world conditions.
Ensure that the infrastructure and hardware match the intended production environment.

Test Cases:
Design test cases that align with the objectives of your AI testing.
Test cases may include input data, expected outputs, and metrics for evaluation.

Example: Imagine you are testing an AI-driven chatbot for customer support in an e-commerce platform. One of the risks you identify is that the chatbot might provide inaccurate product recommendations, potentially resulting in customer dissatisfaction.

Your test planning might include defining objectives such as "assess the chatbot's recommendation accuracy." You would then design test cases that involve various customer queries and evaluate the chatbot's recommendations against expected results.

In the following chapters, we will delve into AI-specific test design and execution techniques, providing practical guidance on how to implement your test strategy effectively. Understanding AI-specific test challenges, risk-based testing, and thorough test planning are essential for a successful AI testing project.

Chapter 6: Test Design and Execution

In this chapter, we will delve into the intricacies of designing and executing tests for AI models. We will explore how to create effective test cases for AI models, the role of test automation, and the importance of continuous testing in AI projects.

Test cases for AI models

Designing test cases for AI models is a critical step to ensure their accuracy and reliability.

Diverse Inputs:

Represent Real Scenarios: Incorporate varied inputs mirroring real-world scenarios.

Example: Test voice recognition systems with diverse accents and languages.

Edge Cases:

Test Extremes: Assess how the model handles inputs at its operational boundaries.

Example: For recommendation systems, test handling of uncommon products.

Adversarial Testing:

Evaluate Responses: Assess model behavior with malicious or unexpected inputs.

Example: Test a language processing model with intentionally incorrect text.

Bias and Fairness Testing:

Detect Discrimination: Design cases to identify bias in model behavior.

Example: Check how a job recommendation system treats different demographic groups.

Realistic Data:

Simulate Actual Usage: Use real-world data in test cases for accurate simulations.

Example: Employ authentic datasets to ensure accurate representation in testing

Here are some essential considerations:

Test automation in AI Testing

Test automation is a valuable tool for AI testing, enabling the efficient execution of test cases.

Here is how it works:

Test Script Generation:

Simulation of User Interactions: Develop automated scripts replicating user actions and data inputs.

Example: Automated scripts simulating varied user interactions with a chatbot AI.

Repetitive Testing:

Consistency Evaluation: Automate repetitive execution of test cases to ensure consistent model performance.

Example: Repeatedly running automated tests to verify image recognition accuracy.

Regression Testing:

Adapting to Model Changes: Automated tests help identify issues resulting from AI model updates or alterations.

Example: Automated regression tests ensuring the stability of a recommender system after modifications.

Scalability:

Efficient Handling of Test Volume: Automation enables the simultaneous execution of a vast number of test cases.

Example: Running multiple automated tests concurrently for load balancing algorithms in AI models.

Example: Consider a machine-learning model that predicts housing prices. You want to test its performance. You design test cases with various inputs, including different property sizes, locations, and amenities. For automation, you create scripts that feed these inputs into the model and check if the predicted prices match your expectations. Automation allows you to execute these tests efficiently and repeatedly as you refine the model.

Continuous testing in AI projects

AI projects often require continuous testing to adapt to changing data and evolving models. Continuous testing involves:

Integration with Development:

Immediate Testing Integration: Incorporate testing into the development pipeline for instant validation of new code or models.

Example: Automated unit tests validating real-time code changes in a recommendation engine.

Monitoring in Production:

Real-time Performance Monitoring: Continuously monitor AI model performance in live environments to promptly identify deviations or issues.

Example: Monitoring a speech recognition system's accuracy in real-time customer interactions.

Retraining and Deployment:

Automated Model Updates: Implement automated retraining and deployment mechanisms for models requiring adaptation.

Example: Automated retraining of a predictive maintenance AI model with new sensor data.

Feedback Loop:

Continuous Improvement Cycle: Establish a feedback loop using production-detected issues to enhance both models and testing procedures.

Example: Using customer feedback to refine sentiment analysis models for social media monitoring.

Continuous testing ensures that your AI model remains effective, reliable, and aligned with the evolving needs of your users and business.

In the following chapters, we will explore techniques for test monitoring, reporting, and addressing issues identified through testing. Effective test design, automation, and continuous testing are essential for maintaining the performance and reliability of AI models over time.

Chapter 7: Test Monitoring and Reporting

In this chapter, we will dive into the vital aspects of test monitoring and reporting for AI models. You will learn how to keep a close eye on your AI models in real-time, identify the key metrics to measure their performance, and effectively report and act upon the findings.

Real-time monitoring of AI models

Monitoring AI models in real-time is crucial to ensuring their ongoing performance and reliability. Here is how to approach it:

Data Drift Detection:

Continuous Monitoring: Monitor incoming data sources regularly for any shifts or alterations in data distribution.

Example: In a stock market prediction model, track changes in market trends, ensuring the model adapts to new patterns.

Model Health Checks:

Regular Assessments: Perform routine evaluations to detect signs of degradation or abnormal behavior in the model.

Example: In an image recognition system, monitor the accuracy rate when identifying specific objects.

Feedback Loops:

User Interaction Analysis: Create systems to collect and analyze user feedback to improve the model's functionality.

Example: For a chatbot, analyze user queries to identify patterns of misunderstood inputs.

Threshold Alerts:

Alerting Systems: Set up automated alerts to signal when the model's performance drops below predefined thresholds.

Example: In a fraud detection system, trigger an alert if there is an increase in false positives.

Metrics for AI testing

Measuring the performance of AI models requires the use of specific metrics. Here are some common metrics for different types of AI models:

Accuracy:

Definition: Represents the ratio of correctly predicted instances to the total instances in the dataset.

Example: In a sentiment analysis model, accuracy measures the proportion of correctly classified sentiments.

Precision and Recall:

Precision: Measures the accuracy of positive predictions among the actual positive cases.

Recall: Measures the proportion of actual positive instances that were correctly predicted.

Use Case: In a medical diagnosis model, precision assesses the accuracy of correctly identifying patients with a specific condition.

F1 Score:

Definition: Harmonic mean of precision and recall, providing a balanced evaluation of model performance.

Application: Used in information retrieval systems to balance precision and recall.

Mean Absolute Error (MAE) and Mean Squared Error (MSE):

MAE: Measures the average of absolute errors between predicted and actual values.

MSE: Measures the average of squared differences between predicted and actual values.

Usage: In predicting housing prices, MAE and MSE assess the deviation of predicted prices from actual prices.

Word Error Rate (WER):

Application: Measures the percentage of incorrectly recognized words in speech recognition systems.

Example: WER evaluates the accuracy of transcriptions in speech-to-text applications.

Bias Metrics:

Purpose: Measures and mitigates potential biases related to sensitive attributes like race or gender.

Use Case: In hiring recommendation systems, bias metrics ensure fairness in the selection process.

Reporting and feedback

Effective reporting and feedback mechanisms are essential for improving

Regular Reports

Objective: Regularly furnish reports detailing the model's performance.

Content: Summarize performance metrics, changes, deviations from benchmarks, and comparisons against defined thresholds.

Issue Tracking

System: Maintain a structured system to track and categorize issues.

Details: Include severity, potential impact, and origin of the issues for comprehensive monitoring.

Root Cause Analysis

Approach: Conduct in-depth root cause analysis upon issue identification.

Focus Areas: Determine if issues stem from data quality, model structure, or integration challenges.

Actionable Insights

Report Recommendations: Reports should not only identify issues but also suggest actionable solutions.

Example: If a recommendation system lacks accuracy, recommendations might involve retraining the model with updated datasets.

User Feedback Integration

Utilization: Actively incorporate user feedback into the feedback and improvement loop.

Benefit: User insights can reveal practical areas for model enhancement and offer real-world perspectives on system performance.

AI models and maintaining user trust:

Example: Imagine you are responsible for monitoring a virtual personal assistant. Real-time monitoring reveals a sharp increase in user dissatisfaction with the assistant's responses. The metrics show a decline in precision and recall for understanding user queries. You analyze user feedback and discover that the assistant is struggling to understand regional accents. You report these findings and recommend implementing an accent-specific training dataset and improving speech recognition models.

In the following chapters, we will continue to explore case studies and practical examples to illustrate how to effectively monitor, report, and improve AI model performance in real-world scenarios. Understanding the importance of these processes is essential for maintaining AI model health and ensuring user satisfaction.

Part 4: AI-Specific Testing Scenarios

Chapter 8: Explainable AI Testing

In this chapter, we will explore the importance of explainable AI (XAI) testing, which focuses on understanding and testing the transparency and interpretability of AI models. We will discuss why interpretability is crucial, methods to test AI model transparency and techniques for conducting explainability testing.

Importance of interpretability

Interpretability is the ability to understand and explain how AI models

Transparency:

Clarity in Decisions: Interpretability enables comprehensible insights into how AI models arrive at specific decisions or predictions.

Understanding: Helps users and stakeholders comprehend the rationale behind the model's outputs.

Trust:

User Confidence: Understanding the model's decision-making process fosters trust in its recommendations.

Acceptance: Users are more likely to accept and embrace model outcomes when the reasoning behind them is clear.

Ethics and Fairness:

Bias Detection: Interpretability aids in detecting biases and discriminatory patterns within AI models.

Rectification: Allows for rectification of biases, ensuring fair and unbiased outcomes.

Regulatory Compliance:

Legal Obligation: Compliance with regulations like GDPR, HIPAA, etc., often mandates AI models to be explainable.

Requirement: Regulatory bodies often require AI models to be interpretable to maintain accountability and transparency.

make decisions. It is critical for several reasons:

Testing AI model transparency

Ensuring the transparency of AI models involves testing and evaluating the model's decision-making processes.

Transparency Metrics:

Defining Metrics: Develop and utilize specific metrics that gauge the clarity and comprehensibility of AI model decisions.

Examples: Metrics could include the percentage of interpretable features or the overall clarity of decision trees within the model.

Black-Box Analysis:

Understanding Black-Box Models: Employ methodologies like LIME or SHAP to interpret and explain predictions made by opaque or complex 'black-box' AI models.

Explanation Generation: These techniques generate explanations on a per-prediction basis, offering insights into individual outcomes.

Visualizations:

Visual Explanations: Develop visual aids such as saliency maps or decision plots to present how the model processes and weighs various input features.

Enhancing Comprehension: These visualizations highlight significant factors influencing predictions in a more understandable format.

Counterfactual Explanations:

What-If Analysis: Provide alternative scenarios illustrating how changes in input data might affect model decisions.

Improving Insights: For instance, a loan rejection might be explained by showcasing what changes in an applicant's data could lead to approval.

Feature Importance Analysis:

Determining Key Factors: Analyze and present the importance of different input features in the decision-making process of the AI model.

Identifying Influential Variables: Helps in understanding which factors hold the most weight in the model's predictions.

Ensuring Transparency in AI Models

Techniques for explainability testing

Here are some techniques to conduct explainability testing:

Feature Attribution Methods

Use techniques like LIME, SHAP, and Integrated Gradients to assess which features contribute most to a model's predictions.

This helps identify the most influential factors.

Model Agnostic Techniques

Apply methods that work across various types of models, providing interpretability regardless of the model's complexity or type. LIME, for instance, is model-agnostic.

Sensitivity Analysis

Test how sensitive the model's predictions are to variations in input data.

This reveals how robust the model is to changes and highlights potential areas for improvement.

Visualization Tools

Utilize visualization tools and libraries to create informative and easily understandable visual representations of model decisions.

Tools like TensorFlow's What-If Tool or Explainable AI libraries can be beneficial.

Example: Suppose you are testing a credit-scoring model. The model decides whether an applicant should receive a loan. You implement feature attribution techniques to understand which factors (income, credit score, age) influence the model's decisions the most. By visualizing the feature's importance, you can provide users with transparent insights into why the model approved or denied a loan application.

In the following chapters, we will dive deeper into case studies and practical examples of explainability testing, highlighting the importance of making AI models transparent and understandable. Understanding and testing AI model interpretability is essential for building trust, ensuring fairness, and complying with regulations in AI applications

Chapter 9: Bias and Fairness Testing

In this chapter, we will focus on the critical aspect of bias and fairness testing in AI. We will explore the importance of addressing bias in AI models, ensuring fairness in AI systems, and strategies to mitigate bias effectively.

Addressing bias in AI models

Bias in AI models occurs when they produce unfair or discriminatory results, often due to biased training data or algorithmic design.

Ethical Responsibility	Legal Compliance
AI models have the power to impact people's lives significantly. It is our ethical responsibility to ensure that they do so fairly and without discrimination.	Many regions have laws and regulations that require AI systems to be fair and non-discriminatory. Violating these laws can lead to legal consequences.
User Trust	**Equity**
Users are more likely to trust and use AI systems that are perceived as unbiased and fair.	Unbiased AI models promote equity by providing equal opportunities and treatment for all users.

Addressing bias is crucial for several reasons:

Ensuring fairness in AI systems

Ensuring fairness in AI systems involves taking a proactive approach to prevent and mitigate bias.

Here is how to achieve it:

Data Evaluation

Thoroughly evaluate training data to identify potential sources of bias.

Check for underrepresented groups, unbalanced data, and data collection processes that may introduce bias.

Fair Data Sampling

If you detect bias in your data, consider oversampling underrepresented groups, generating synthetic data, or using data augmentation to balance your dataset.

Algorithmic Fairness

Use fairness-aware algorithms that explicitly consider fairness constraints during training.

These algorithms aim to reduce disparate impacts on different groups.

Transparency

Make the AI model's decision-making process transparent and interpretable.

Explain how the model reached its decisions to build user trust.

Regular Auditing

Continuously audit the AI model's performance for bias and fairness.

Monitor its behavior in real-world scenarios and make adjustments as needed.

Bias mitigation strategies

Mitigating bias in AI models requires proactive strategies.

Reweighting

Assign different weights to training examples to give more importance to underrepresented groups.

This can help balance the model's performance.

Adversarial Debiasing

Introduce an adversarial network that aims to minimize the influence of sensitive features in the model's decision-making process.

Fair Representations

Transform the data to create fair representations, reducing the influence of sensitive attributes while retaining essential information.

Calibration

Calibrate the model's output to ensure that it is equally accurate across different groups.

Post-Hoc Analysis

Conduct post-hoc analysis to identify and rectify bias in model predictions.

This analysis can involve techniques like re-ranking or re-weighting model outputs.

Here are some techniques to consider:

Example: Imagine you are working on a recruitment AI system that helps filter job applicants. During testing, you discover that the system tends to reject candidates with names from underrepresented minority groups at a higher rate than others. To address this bias, you reweight the training data, giving more significance to underrepresented names. Additionally, you introduce fairness-aware algorithms that explicitly consider ethnicity as a fairness constraint during model training. These measures help reduce the discriminatory impact on job applicants.

In the following chapters, we will delve into practical case studies and examples of bias and fairness testing, emphasizing the significance of building AI systems that treat all users equitably. Understanding and effectively addressing bias is essential for creating AI systems that are trustworthy, ethical, and legally compliant.

Chapter 10: Security and Ethical AI Testing

In this chapter, we will explore the vital aspects of security and ethical AI testing. We will delve into the security challenges in AI, ethical considerations that should guide your testing, and the importance of complying with relevant regulations.

Security challenges in AI

AI systems can introduce security vulnerabilities and risks, making security testing a crucial aspect of AI development. Here are some key security challenges in AI:

Data Privacy

AI models often require sensitive or personal data for training.

Protecting this data from unauthorized access and breaches is essential.

Adversarial Attacks

AI models can be vulnerable to adversarial attacks, where attackers manipulate inputs to deceive the model into making incorrect predictions.

For example, in image recognition, an attacker could add imperceptible noise to an image to mislead the model.

Model Security

Protecting the integrity of the AI model itself is crucial.

Unauthorized modifications to the model can result in malicious behavior or incorrect predictions.

Explainability

Non-transparent AI models can be exploited by attackers to manipulate the model without detection.

Understanding the model's decision-making process is vital for detecting and preventing such attacks.

Ethical considerations in AI testing

Ethical considerations play a significant role in AI testing. Ensuring ethical AI is essential for user trust, fairness, and legal compliance.

Here is how to approach ethical AI testing:

Bias and Fairness

Test AI models for bias and fairness to ensure that they treat all users equitably.

Mitigate bias and ensure that sensitive attributes do not lead to discrimination in model predictions.

Privacy

Safeguard user privacy by testing how the AI system handles personal data.

Ensure that data anonymization and encryption techniques are in place to protect user information.

Transparency

Promote transparency by testing how well the AI model's decisions can be understood and explained.

Transparent models are less likely to engage in unethical or harmful behavior.

Accountability

Test and implement mechanisms for AI model accountability.

Ensure that there is a clear path to identifying responsible parties if the AI system makes harmful or unethical decisions.

Compliance with regulations

Compliance with regulations is critical for AI systems, as many regions have introduced laws to ensure the responsible development and use of AI. Key regulations to be aware of include:

<table>
<tr>
<td>

GDPR (General Data Protection Regulation)

GDPR places strict requirements on how personal data is handled, including AI systems that process personal data.

</td>
<td>

HIPAA (Health Insurance Portability and Accountability Act)

HIPAA regulates the handling of healthcare data, making it crucial for AI systems in the healthcare sector.

</td>
</tr>
</table>

AI Ethics Guidelines

Some regions, such as the European Union, have introduced AI ethics guidelines to ensure AI systems are developed and used responsibly.

Example: Consider an AI-driven healthcare diagnostic system. During testing, you discover that the system's algorithm is inadvertently biased against patients from certain ethnic backgrounds, leading to incorrect diagnoses. To address this ethical concern, you retrain the model with a more diverse dataset that represents a wider range of patients. You also ensure that the system complies with HIPAA regulations by protecting patient data and maintaining privacy.

In the following chapters, we will delve into real-world case studies and practical examples of security and ethical AI testing, emphasizing the importance of ethical and secure AI development. Understanding and addressing security challenges and ethical considerations is essential for building AI systems that are trustworthy, respectful of privacy, and compliant with regulations.

Part 5: Case Studies and Practical Examples

Chapter 11: Real-world AI Testing Projects

Case studies from industry

Autonomous Vehicle Testing

An automotive company developed a self-driving car powered by AI. Testing involved a combination of simulation, controlled environments, and on-road testing.

Challenges included ensuring the AI system's response to diverse road conditions and the accuracy of perception and decision-making modules.

Lessons learned emphasized the importance of extensive simulation testing to cover rare scenarios and frequent validation of sensor inputs in real-world conditions.

Healthcare AI

A medical imaging company deployed an AI system to assist radiologists in identifying anomalies in medical images.

Testing focused on verifying the system's accuracy and its adherence to patient privacy regulations.

Challenges included addressing data privacy concerns and ensuring the model's generalizability across different hospitals and patient populations.

Best practices involved data anonymization and continuous monitoring of model performance.

Natural Language Processing (NLP) Chatbot

A customer service organization implemented an NLP chatbot to handle customer inquiries.

Testing involved validating the bot's understanding of user queries, its ability to provide relevant responses, and its performance in handling various languages and accents.

Challenges included addressing bias in language models and ensuring the bot's adaptability to new customer queries.

Lessons learned emphasized the need for regular model retraining and addressing fairness issues in NLP.

Lessons learned and best practices

Test Diversity

Real-world AI testing should encompass a diverse set of test cases and scenarios, representing a wide range of real-world conditions.

Comprehensive test coverage helps identify and address potential issues early.

Bias Mitigation

Addressing bias is a critical aspect of AI testing.

Regularly assess the model for bias, develop strategies for bias mitigation, and ensure fairness in AI systems.

Continuous Testing

Implement continuous testing practices to adapt to evolving AI models and data.

This involves monitoring model performance in production and conducting ongoing validation.

Data Quality

Maintain high data quality standards. Ensure that training data is clean, diverse, and accurately represents the target user base.

Privacy and Compliance:

Be vigilant about data privacy and legal compliance, especially in AI applications that involve sensitive information.

Regularly audit data handling practices to maintain compliance.

Human-in-the-Loop Testing

Consider involving humans in the testing process, especially when evaluating AI systems that interact with users.

Human-in-the-loop testing can uncover usability issues and provide valuable feedback.

Ethical Considerations

Prioritize ethical considerations in AI testing.

Ensure that AI systems treat all users fairly and equitably.

Be prepared to address ethical dilemmas that may arise during testing.

Transparent Documentation

Maintain transparent documentation of test cases, results, and any issues identified during testing.

Clear documentation aids in troubleshooting and improvement efforts.

Example: In the autonomous vehicle case study, a valuable lesson learned was the importance of simulating rare and challenging scenarios, such as extreme weather conditions and complex traffic situations. As a best practice, the company established a continuous testing process that involved regularly updating the simulation environment with new scenarios. This proactive approach allowed them to identify and address potential issues before they occurred in real world testing, enhancing the safety and reliability of their AI-driven vehicles.

Real-world AI testing projects often involve dynamic challenges that require a combination of technical skills, ethical considerations, and adaptability. Learning from these case studies and following best practices can help you navigate the complexities of AI testing and contribute to the successful deployment of AI systems in various industries.

Chapter 12: Hands-on Examples

In this chapter, we will provide you with practical, hands-on examples for AI testing. These exercises include code snippets and testing scenarios to help you gain real-world experience and improve your skills in AI testing.

Example 1: Testing a Machine Learning Model

Let us imagine you are tasked with testing a machine-learning model for sentiment analysis. This model takes a text input and predicts whether the sentiment of the text is positive, negative, or neutral. Here is a testing scenario:

Scenario: You are testing a sentiment analysis model, and you need to assess its accuracy. You have a dataset of 1,000 customer reviews. For each review, the model should predict the sentiment, and you will

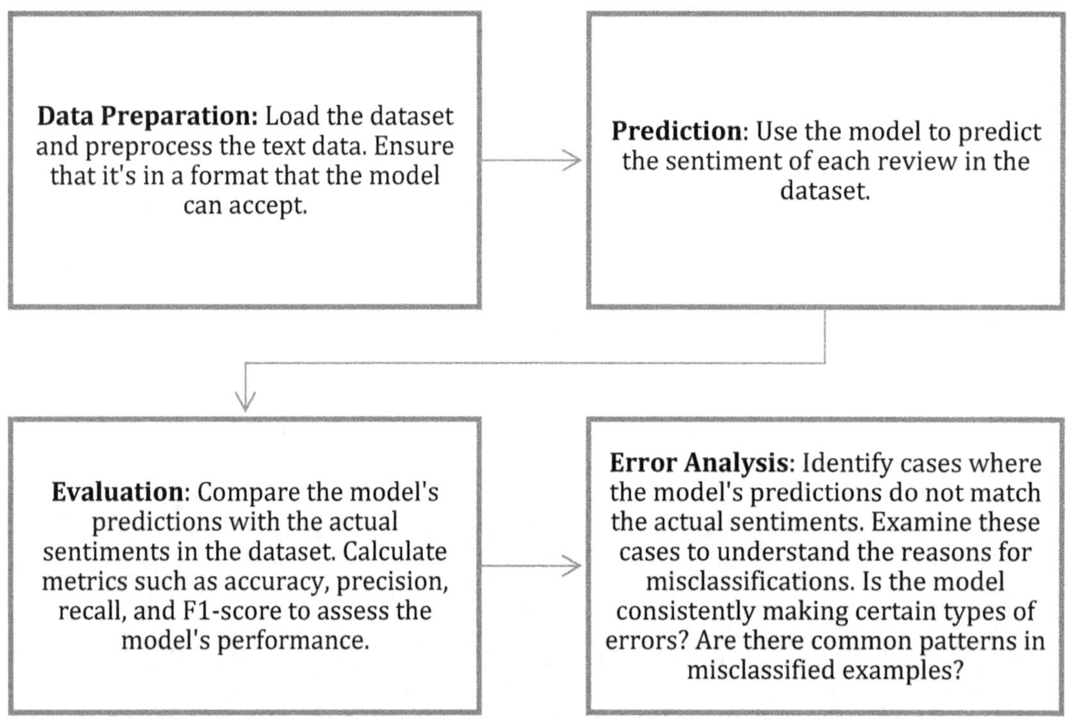

compare these predictions to the actual sentiments in the dataset.

Example 2: Testing an AI Chatbot

Let us explore a scenario where you are testing an AI chatbot designed to assist customers with common inquiries. The chatbot should provide accurate responses to user queries. Here is a testing scenario:

Scenario: You are testing an AI chatbot for a customer service application. The chatbot should understand and respond to user queries effectively. You have a set of test cases representing different types of customer inquiries.

Steps:

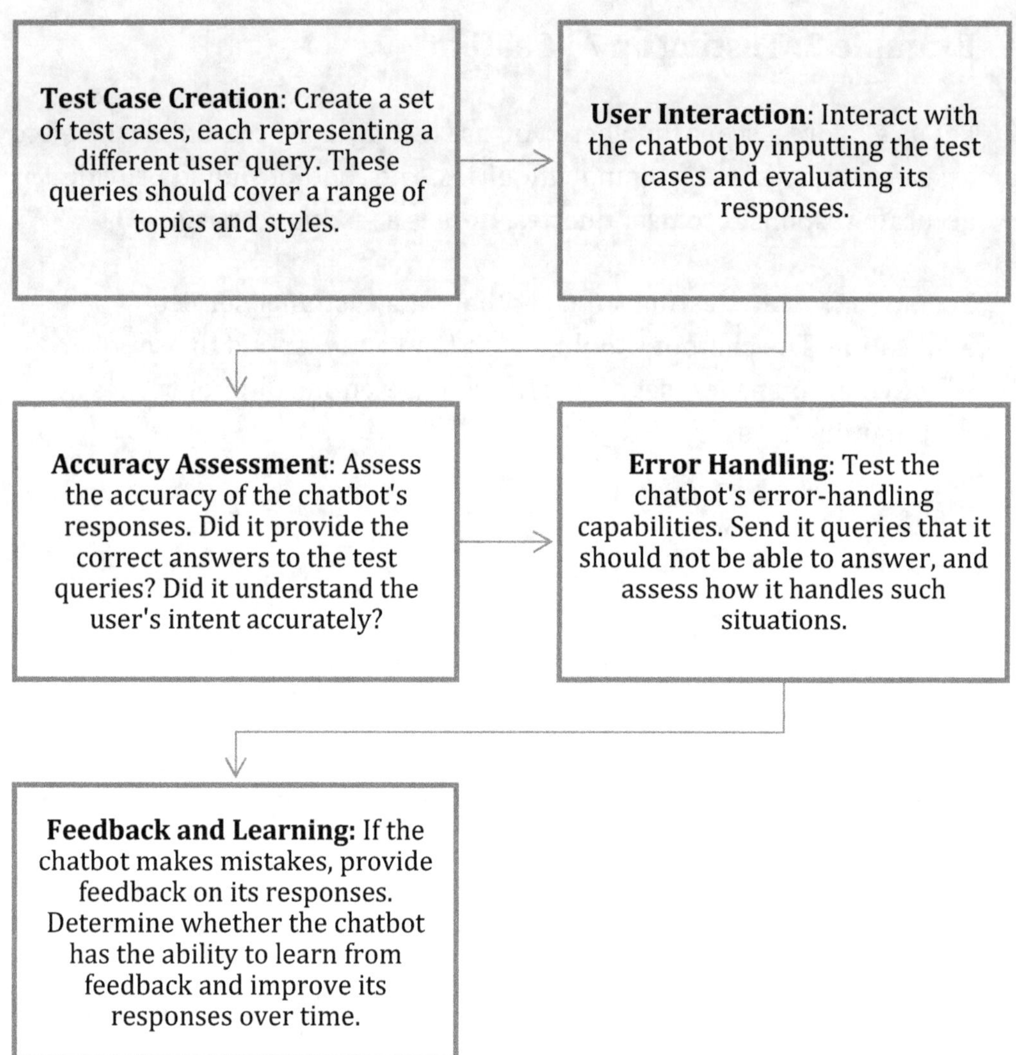

By engaging in these practical exercises, you will gain valuable experience in testing AI models and systems, which is essential for becoming a proficient AI tester. The ability to create, execute, and evaluate test scenarios in a hands-on manner will enhance your skills and readiness for real-world AI testing projects.

Part 6: The Future of AI Testing

Chapter 13: Emerging Trends in AI Testing

In this chapter, we will explore the emerging trends in AI testing, focusing on how artificial intelligence is reshaping different aspects of software testing. We will discuss AI in DevOps, AI in test automation, and AI in security testing, providing examples of how these trends are impacting the industry.

Emerging Trend

AI is increasingly integrated into DevOps practices to enhance the speed and quality of software development and testing.

Impact

AI in DevOps helps teams automate repetitive tasks, detect and predict issues in real-time, and optimize the deployment process.

Example

Consider a DevOps team that uses AI to predict potential bottlenecks in their continuous integration and continuous deployment (CI/CD) pipeline.

The AI system analyzes historical data on build times, test failures, and deployment success rates.

It then provides recommendations on how to optimize the pipeline to reduce build times and prevent common deployment failures.

This proactive approach minimizes delays and enhances the overall efficiency of the software development process.

AI in DevOps

Emerging Trend

AI-driven test automation leverages machine learning to create, maintain, and execute test scripts efficiently.

Impact

AI in test automation enhances test coverage, accelerates test execution, and reduces maintenance efforts by automatically adapting to application changes.

Example

A test automation framework employs AI to self-heal test scripts.

When the application's user interface undergoes minor changes, traditional test scripts might break.

AI-driven automation can recognize these changes and automatically adjust the test scripts to match the new UI elements.

This adaptive capability reduces the need for manual script updates and improves the robustness of automated tests.

AI in test automation

Emerging Trend

AI is playing a crucial role in enhancing security testing by automating vulnerability detection, threat analysis, and risk assessment.

Impact

AI-driven security testing identifies vulnerabilities and threats more comprehensively and rapidly than manual testing.

It helps organizations proactively secure their applications and data.

Example

A security testing tool powered by AI performs continuous scanning of a web application to identify potential vulnerabilities and threats.

When the tool detects an unusual pattern of behavior that may indicate an attack, it triggers alerts and provides a detailed analysis of the potential threat.

This real-time monitoring and threat analysis enable organizations to respond quickly to security breaches and mitigate risks effectively

AI in security testing

By embracing these emerging trends in AI testing, organizations can streamline their software development processes, improve the quality of their products, and enhance their security posture. As AI continues to evolve, its impact on software testing is expected to grow, making it essential for testers and quality assurance professionals to stay updated with these trends to remain competitive in the industry.

Chapter 14: The Role of AI Testers in Industry

Career opportunities for AI testers

The field of AI testing is rapidly expanding, and AI testers are in high demand across various industries. Here are some career opportunities for AI testers:

AI Test Engineer

As an AI test engineer, your primary responsibility is to design and execute tests for AI models, ensuring their accuracy, performance, and reliability.

You'll work on testing various AI applications, from chatbots to autonomous vehicles.

Data Quality Analyst

Data is the lifeblood of AI systems. Data quality analysts focus on ensuring that the data used to train AI models is accurate, representative, and free from bias.

They play a critical role in mitigating bias in AI.

AI Test Automation Engineer

These professionals specialize in creating automated test scripts and frameworks for AI applications.

They use tools and frameworks to automate the testing process and increase testing efficiency.

AI Test Manager

AI test managers oversee testing teams and projects.

They are responsible for test planning, resource allocation, and ensuring the quality of AI testing efforts.

They often interact with other stakeholders in the development process.

AI Security Tester

AI security testers specialize in evaluating the security aspects of AI applications.

They identify vulnerabilities and potential threats in AI systems to ensure data and user privacy.

Skill development and specialization

To excel in the role of an AI tester and stand out in the industry, it's essential to develop specific skills and consider areas of specialization:

Machine Learning Knowledge

Gain a deep understanding of machine learning concepts, algorithms, and frameworks.

Familiarize yourself with supervised learning, unsupervised learning, and deep learning.

Testing Tools and Frameworks

Learn to work with AI testing tools and frameworks.

Familiarize yourself with popular AI testing tools such as TensorFlow, PyTorch, and specialized testing frameworks.

Data Analysis Skills

Develop strong data analysis skills to assess data quality, identify patterns, and understand the data used to train AI models.

Bias Mitigation

Specialize in bias mitigation techniques. Understand how to identify and address bias in AI models, especially in applications like natural language processing.

Automation Expertise

If you choose to specialize in automation, become proficient in test automation tools and frameworks for AI applications.

This includes scripting, test case design, and continuous testing.

Ethical Considerations

Stay informed about ethical considerations in AI testing.

Develop a strong ethical framework to ensure that AI systems are fair, transparent, and respectful of user privacy.

Example: Let's say you're an AI test engineer specializing in natural language processing (NLP). You have developed expertise in using NLP models and frameworks like spaCy and NLTK. In a recent project, you were responsible for testing a chatbot's language understanding and generation capabilities. You created test cases that assessed the chatbot's ability to handle a variety of user queries. You also implemented bias detection techniques to ensure that the chatbot's responses were unbiased and fair, especially when dealing with sensitive topics. Your specialization in NLP and bias mitigation makes you a valuable asset in the AI testing field.

As AI continues to evolve and integrate into various industries, AI testers will play a crucial role in ensuring the reliability, fairness, and security of AI applications. Developing the right skills and considering areas of specialization will enable you to thrive in this dynamic and rewarding field.

Part 7: Conclusion and Final Thoughts

Part 7: Conclusion and Final Thoughts

Chapter 15: Conclusion

In this final chapter, we will reflect on the importance of AI testing and the exciting journey ahead for AI testers. We will highlight the critical role AI testers play in the world of artificial intelligence and share insights into what the future may hold.

The importance of AI testing

AI testing is a fundamental and indispensable component of AI development. It ensures that AI systems and applications are reliable, safe, ethical, and perform as intended. Here is why AI testing is of

User Trust

AI testers are responsible for building and maintaining user trust. When users interact with AI-driven systems, they expect accurate and reliable results.

Testing is essential to meet these expectations.

Safety and Security

AI applications, especially those in healthcare, autonomous vehicles, and finance, have significant safety and security implications.

AI testing helps identify vulnerabilities and ensures the security of these systems.

Ethical Considerations

Testing is crucial for addressing ethical concerns, such as bias and fairness.

AI testers play a pivotal role in detecting and mitigating bias, ensuring that AI systems treat all users fairly and equitably.

Compliance

AI applications often need to adhere to legal and regulatory requirements.

AI testers ensure that AI systems comply with data privacy laws, industry regulations, and ethical guidelines.

paramount importance:

The journey ahead for AI testers

The journey for AI testers is filled with exciting opportunities and challenges.

Advancements in AI Testing Tools

AI testing tools and frameworks will continue to evolve, offering testers more powerful and efficient ways to evaluate AI applications.

Specialization Opportunities

AI testers will have the opportunity to specialize in various domains, from NLP and computer vision to autonomous systems and healthcare AI.

AI-driven Test Automation

The role of AI in test automation will expand, allowing testers to automate repetitive tasks, adapt to changes in applications, and accelerate testing processes.

Continuous Learning

AI is a rapidly evolving field. AI testers will need to commit to continuous learning to stay current with the latest developments and best practices.

Ethical Leadership

AI testers will take on ethical leadership roles, ensuring that AI systems are developed and used responsibly and ethically.

Here is what the future may hold:

Example: Consider an AI tester who specializes in autonomous vehicle testing. As the field of autonomous vehicles grows, their role becomes increasingly critical. They work on testing the AI algorithms that enable self-driving cars to make decisions on the road. This includes creating scenarios for testing the vehicle's ability to respond to various situations, from traffic jams to unexpected obstacles. The tester plays a pivotal role in ensuring the safety of autonomous vehicles and building public trust in this revolutionary technology.

In conclusion, AI testing is a dynamic and essential field that will continue to shape the future of AI applications. AI testers are at the forefront of this transformation, ensuring that AI systems are reliable, secure, and ethical. As AI continues to evolve, the journey ahead for AI testers is filled with opportunities to make a profound impact on the technology landscape and contribute to the responsible development and deployment of AI systems.

Chapter 16: Appendices

In this final chapter, we provide valuable appendices to support your understanding of AI testing. These appendices include a glossary of AI and testing terms to clarify important terminology and a list of useful resources and references for further exploration.

Appendix A: Glossary of AI and Testing Terms

This glossary contains key terms related to AI and software testing, helping you navigate the field and understand the terminology used throughout this book.

Artificial Intelligence (AI):

AI refers to the simulation of human intelligence in machines that are programmed to mimic human actions and cognitive functions.

Machine Learning:

A subset of AI that enables systems to learn and improve from experience without being explicitly programmed.

Deep Learning:

A type of machine learning that employs neural networks with multiple layers, capable of learning representations of data with increasing complexity.

Supervised Learning:

A machine learning technique where models are trained on labeled data, enabling them to make predictions or decisions.

Unsupervised Learning:

A machine learning technique that involves training models on unlabeled data, allowing them to identify patterns or structures independently.

Reinforcement Learning:

A type of machine learning where an agent learns to make decisions through trial and error, receiving feedback in the form of rewards or penalties.

Narrow AI (Weak AI):

AI systems designed and trained for specific tasks, lacking the broad capabilities of human-like intellect.

General AI (Strong AI):

AI systems capable of understanding, reasoning, and performing tasks across a wide range of domains, akin to human intelligence.

Algorithm:

A step-by-step procedure or set of rules to solve problems or perform tasks.

Testing:

The process of evaluating a system's functionalities or behavior to identify errors or gaps.

Test Automation:

The use of software tools and scripts to automate the execution of test cases and reporting of results.

Regression Testing:

Re-running tests to ensure changes in the codebase do not adversely affect previously working functionalities.

Black-Box Testing:

Testing a system's functionality without knowing its internal structure or codebase.

White-Box Testing:

Testing that involves assessing the internal structure and logic of a system's code.

Test Case:

A set of conditions or variables under which a tester determines whether an application, feature, or functionality works as intended.

Validation:

The process of ensuring that the developed software meets the specified requirements.

Verification:

The process of evaluating whether software complies with its specified design and requirements.

Glossary:

An alphabetical list of terms, along with their meanings or explanations.

AI Ethics:

The field of study concerned with ensuring AI systems operate ethically and responsibly, addressing issues of bias, fairness, and accountability.

Bias in AI:

The inclination of AI systems to produce results that are systematically prejudiced or unfair.

Overfitting:

Occurs when a machine learning model learns the training data excessively, resulting in poor performance when faced with new, unseen data.

Underfitting:

Happens when a machine learning model is too simple to capture the underlying patterns in the training data, leading to inaccurate predictions.

Feature Engineering:

The process of selecting, transforming, or creating new features from raw data to enhance model performance in machine learning.

Hyperparameters:

Parameters set before training a machine learning model that determine its structure or learning process (e.g., learning rate, depth of a decision tree).

Cross-Validation:

A technique used to assess the performance of a machine learning model by partitioning the dataset into subsets for training and testing.

Ensemble Learning:

A machine learning technique that combines predictions from multiple models to improve overall performance and accuracy.

Bias-Variance Tradeoff:

The balance between model complexity (variance) and its ability to capture the underlying patterns (bias) in machine learning.

Precision:

In classification, precision measures the ratio of correctly predicted positive observations to the total predicted positives.

Recall:

In classification, recall measures the ratio of correctly predicted positive observations to the actual positives in the dataset.

F1 Score:

A metric used in classification that combines precision and recall, providing a balance between the two.

ROC Curve (Receiver Operating Characteristic Curve):

A graphical representation that illustrates the performance of a binary classifier across various thresholds.

AUC (Area Under the Curve):

The area under the ROC curve, which quantifies the classifier's ability to distinguish between classes.

Confusion Matrix:

A table that visualizes the performance of a classification model, showing correct and incorrect predictions.

Feature Importance:

Indicates the significance of each feature in influencing predictions made by a machine learning model.

Anomaly Detection:

The process of identifying patterns or instances in data that deviate significantly from the norm or expected behavior.

Model Deployment:

The process of integrating a trained machine learning model into a production environment for real-time use.

Natural Language Understanding (NLU):

A subset of natural language processing focused on comprehending and interpreting human language.

Tokenization:

The process of breaking text into smaller units called tokens, often words or phrases, for analysis in natural language processing.

Named Entity Recognition (NER):

The identification and classification of named entities such as names, organizations, or locations within text.

Transfer Learning:

A machine learning technique where knowledge gained from training on one task is applied to a different but related task, often using pre-trained models.

Appendix B: Useful Resources and References

This section provides a curated list of resources and references to further your knowledge and expertise in AI testing. These resources include books, online courses, websites, and industry publications.

Books:
a) "Hands-On Machine Learning with Scikit-Learn, Keras, and TensorFlow" by Aurélien Géron.
b) "AI and Machine Learning for Business" by Scott M. Graffius.

Online Courses:
a) Coursera (coursera.org): Offers a wide range of AI and machine learning courses from top universities and institutions.
b) edX (edx.org): Provides AI courses, including the AI and Machine Learning for Business Professional Certificate.

Websites:
a) ISTQB (istqb.org): The official website of the International Software Testing Qualifications Board provides information about ISTQB certifications.
b) Towards Data Science (towardsdatascience.com): A platform on Medium that offers articles and tutorials on AI and machine learning.

Industry Publications:
a) "AI in Testing" by Gartner (gartner.com): Provides research and insights on AI trends in software testing.
b) "Journal of Software Testing, Verification and Reliability" (onlinelibrary.wiley.com/journal/2517-1486): A journal covering research in software testing.

These appendices are designed to enhance your understanding and provide a valuable reference as you continue to explore the world of AI testing. Whether you're looking to clarify specific terms or delve deeper into AI testing resources, these appendices are a valuable addition to your knowledge toolkit.